Vilson Gruber
Roderval Marcelino
Luan Casagrande

IOT and Industry 4.0

Vilson Gruber
Roderval Marcelino
Luan Casagrande

IOT and Industry 4.0

Fundamentals and application in education

ScienciaScripts

Imprint
Any brand names and product names mentioned in this book are subject to trademark, brand or patent protection and are trademarks or registered trademarks of their respective holders. The use of brand names, product names, common names, trade names, product descriptions etc. even without a particular marking in this work is in no way to be construed to mean that such names may be regarded as unrestricted in respect of trademark and brand protection legislation and could thus be used by anyone.

Cover image: www.ingimage.com

This book is a translation from the original published under ISBN 978-613-9-61604-6.

Publisher:
Sciencia Scripts
is a trademark of
Dodo Books Indian Ocean Ltd. and OmniScriptum S.R.L publishing group

120 High Road, East Finchley, London, N2 9ED, United Kingdom
Str. Armeneasca 28/1, office 1, Chisinau MD-2012, Republic of Moldova, Europe
Printed at: see last page
ISBN: 978-620-7-87571-9

Table of contents:

Authors:
Vilson Gruber
Roderval Marcelino
Lucas Boeira Michels
Sarah Resende Guerra
Luan Carlos Casagrande
Rodrigo Cesar Nunes Maciel

IOT AND INDUSTRY 4.0: FUNDAMENTALS AND APPLICATION IN THE EDUCATION

Editors:
Vilson Gruber
Luan Carlos Casagrande
Rodrigo Cesar Nunes Maciel
Brazil - 2018

Federal University of Santa Catarina - UFSC Campus Araranguá/SC
Telecommunications Laboratory - LABTEL/UFSC/ Araranguá/SC/Brazil

The world is a book, and those who sit at home read only one page (St. Augustine).

Chapter 1

1 INTRODUCTION

As a result of technological advances, the internet has become one of the most widespread technological means in recent years. As a result of this technological revolution, the *Internet of Things* (IoT), which is a technology that integrates ubiquitous computing, has emerged to incorporate electronic objects into the Internet, with the aim of offering users interconnected devices that can capture information and use it intelligently. Gartner (2014) points out that the estimated number of connected things in use in 2015 is 4.9 billion, 30% more than in 2014. And the outlook for 2020 is that the use of connected things could reach 25 billion.

According to Kopetz (2011), the Internet was initially a small research network made up of a few nodes and in the last 50 years, it has become a communication network that more than a billion users use around the world.

As a result of the Internet revolution, according to Kopetz (2011), a new context has emerged as a result of the reduction in the cost and size of electronic devices. The Internet can provide a connection so that ordinary physical objects, with the addition of small electronic devices that link the real world to the world of information, can acquire local intelligence. In this context, Kopetz (2011, p. 308) conceptualizes this new artifice as: "A smart object is thus a physical cybernetic system or an embedded system, consisting of a thing (the physical entity) and a component (computer), which processes sensor data and supports a wireless communication link to the Internet".

Hosain (2015) describes the Internet of Things as a comprehensive new technology in which the results will be beneficial for companies in terms of expanding the business market for current companies and creating new companies to meet the demand of this new technological segment.

Regarding the technologies needed for IoT to be widely used, Vázquez (2013) explains that Wi-Fi networks present in many places or through 2G, 3G and 4G telephone systems allow objects to remain connected to the internet with services that promote knowledge and intelligence.

With the expansion of the IoT, according to Gartner (2014), the economic impact of the Internet of Things will involve consumers, companies, government authorities, hospitals, among other sectors that must adapt to the use of this new technology in order to acquire the benefits of the Internet of Things. For this new technological scenario, it is estimated that spending in 2015 is around US$ 69.5 billion and the estimated spending for 2020 is US$ 263 billion.

In addition, the operational management of companies, according to Hosain (2015), can be improved through the use of the IoT and M2M (*machine to machine*) communication, as the operations carried out by these companies will be seen in greater detail than before. The use of the IoT with M2M and sensors integrated into machines and processes would result in a profit of $1.95 trillion between 2015 and 2022.

Gartner (2014) also predicts that the manufacturing, utilities and transportation sectors that make use of the Internet of Things will together have

736 million connected things by 2015.

The expectation for 2020 is that 1.7 billion Internet of Things units will be installed, with utilities coming first, manufacturing second and government third.

In addition, Gartner (2014) points out that consumers will be responsible for the applications targeted and used by consumers will govern the number of connected things. At the other end of the spectrum, companies will be responsible for the largest percentage of expenditure.

Table 1 shows Gartner's estimates for the base units of the Internet of Things, i.e. the categories in which the Internet of Things will have the greatest growth impact in relation to connected things:

Table 1: Internet of Things according to installed base units by Category [1]

Categories	**2013**	**2014**	**2015**	**2020**
Automotive	96.0	189.6	372.3	3,511.1
Consumer	1,842.1	2,244.5	2,874.9	13,172.5
Generic Business	395.2	479.4	623.9	5,158.6
Vertical Business	698.7	836.5	1,009.4	3,164.4

1 Source: GARTNER, 2014 apud GARTNER, INC, 2015.

Chapter 2

2 IOT CONCEPTS

Like other developments, IoT is the result of the accelerated evolution of technology and the search for interconnection between computer components with the aim of transforming the lives of people and companies by offering convenience and intelligent use of information, thus enabling more efficient communication, more effective, productive and profitable processes in various areas.

According to Vázquez (2013, p.64):

> The term Internet of Things was proposed by Kevin Ashton in 1999, in a presentation that defended the idea that RFID tags associated with physical objects would give them an identity under which they could generate data about themselves or what they perceived and publish it on the Internet.

Gubbia (2013, p. 1647) emphasizes that the internet of things was defined by the RFID group as: "The worldwide network of objects interconnected in a uniquely addressable way based on conventional communication protocols". In addition to the term coined by Kevin Ashton, Vázquez (2013) points out that the concept of the internet of things is based on the principle that the objects around us will be the main characters on the internet, creators and consumers of information produced by themselves, systems or other people. On the other hand, Gubbia (2013) conceptualizes the internet of things as interconnected sensors and actuating devices that encourage the sharing of information between platforms using a unified structure, an environment designed to facilitate innovative applications. This is made possible by using cloud computing, which serves as a unifying element for data analysis and information representation, and by seamless ubiquitous sensing.

For Kopetz (2011), the perspective of the Internet of Things emphasizes that ordinary things that are part of our environment have their own unique identity, absorb information from various sources and gain intelligence so that an intelligent world can develop. The Internet of Things will determine a uniform access standard for real-world objects, so that this intelligent world benefits the more effective operation of the global economy and support systems. There will also be a distinction in power between citizens who have access to and control of information and those who do not.

Internet-connected objects, says Vázquez (2013), have a relevant characteristic in common based on the different types of sensors: making the invisible visible, i.e. the data will be transmitted because it has always existed but has never been accessed. In addition, the author adds that another important factor for the success of smart objects is social collaboration to provide useful information collected from all users.

Similarly, according to Kopetz (2011), communication via a *wireless* network makes it easier to connect to a smart object without using a physical network, where smart objects can move around the real environment without losing their characteristics. One example is the Global Positioning System (GPS), where the variety of signals available makes this intelligent object provide its location and time services, according to the needs and situation at the time.

Chapter 3

3 TECHNOLOGIES

The IoT is consolidating and delivering results as a result of the integration of the technologies used, whether they are related to hardware (electronic components), network architecture, data transmission and storage, applications and/or new technology trends that are incorporated into this comprehensive technology. In fact, according to Vázquez (2013), the combination of the Internet of Things with new trends such as *Big Data* and open data is creating a field for a new analytical service capable of making connections between elements considered to be distant.

In addition, Kopetz (2011) points out that *RFID* (*Radio Frequency Identification*) technology, the predecessor of the IoT, is applied commercially in the area of logistics, where the numerous benefits have led to the development of *RFID tags* and readers for use in the management of goods, where it is possible to control movement in real time, control stock more effectively, manage shelf space and, above all, human resources to carry out this management will be reduced considerably. On the other hand, some technical problems related to the Internet of Things are highlighted by Kopetz (2011):

- **Internet integration**: this action, according to energy and components

This can be done indirectly (limited energy) using a central station connected to the internet which acts as a *web* server providing access to smart objects. A number of companies are working together to develop solutions and standards aimed at integrating smart objects directly into the Internet. The *IETF* (*Internet Engineering Task Force*) has set up a working group to find a solution and integrate the *IPV6* standard with the *IEEE* 802.15.4 wireless communication standard of the nearby network. Normally, the security of these systems using the Internet of Things will be carried out by a firewall which will have the task of preventing intruders from controlling the smart object;

- **Naming and Identifying**: a naming architecture is required

to identify an intelligent object so that an access route can be determined. In order to do this, it is essential that the naming is in accordance with the context to generate a unique and universally accepted nomenclature. The EPC (*Electronic Product Code*), used by the *RFID* community, exemplifies a model for identifying smart objects where the optical barcode is unique to each class of object;

- ***Near Field Communication*:** in order to use smart objects in a small area, the Internet of Things requires *WLANs* (*Wireless Local Area Networks*) and *WPANs* (*Wireless Personal Area Networks*). To this end, the *IEEE* 802.15 standard has developed *WPAN network* standards, including the Bluetooh and ZigBee networks compatible with the 802.15 standards.

As Vázquez (2013) points out, the Internet of Things is encouraging new projects and ideas, because the technologies used to develop connected objects do not require periods of training or technical knowledge and skills. Among these technologies is Arduino, a low-cost platform combined with simplified programming, which facilitates learning and encourages the development of projects using faster prototyping cycles and the availability of its codes. It is an open-source technology that encourages interaction and the exchange of knowledge between developers.

3.1 IOT COMPONENTS

Since the IoT is a pervasive computing technology in which various elements are integrated to develop intelligent objects that perform functions in various areas, the structure of the IoT requires certain items and resources.

In the view of Gubbi et al (2013), three components of the Internet of Things are necessary for perfect ubiquitous computing:

a) *Hardware* - sensors, actuators and embedded communication hardware make up the communication *hardware;*
b) *Middleware* - storage and computing tools for data analysis;
c) Presentation - to understand the visualization and interpretation tools that can be accessed on different platforms and available for different applications.

There are also some simplifying technologies that form part of the Internet of Things components, as described below:

d) ***RFID (Radio Frequency Identification):*** *RFID* is considered to be the technology that has contributed to the evolution of the embedded communication model, where wireless data communication is possible through the design of microchips that help to automatically identify data (JUELS, 2006; WELBOURNE, 2009 apud GUBBI et al, 2013).
e) ***Wireless Sensor Networks (WSN):*** the sensor shares its data between sensor nodes and forwards it to a distributed or centralized system for analysis. The *WSN* monitoring network is made up of the following components:
    - **WSN hardware** - A node (*WSN hardware* core*),* consisting of sensor interfaces, processing and transceiver units, and power supply. They usually have several A/D converters for the sensor interface (ATZORI, 2010 apud GUBBI et al, 2013);
    - ***WSN* communication stack** - Nodes must communicate with each other to transmit data in one or more stages to a base. The communication stack to the sink node must interact with the outside world via the Internet as a gateway to the *WSN* subnet and the Internet (GHOSH, DAS, 2008 apud GUBBI et al, 2013);

- ***WSN Middleware - A*** device that associates cyberinfrastructure with a SOA (Service Oriented Architecture) and sensor networks that, in an independent and established manner, offers access to heterogeneous sensor resources (GHOSH, DAS, 2008 apud GUBBI et al, 2013);
- **Secure data aggregation** - In order to extend the lifetime of the network and guarantee reliable data captured by the sensors, an effective and secure resource is needed to aggregate this data. The WSN (Wireless Sensor Network) has the characteristic of node failures, so it must be able to restore itself. Ensuring security is essential to avoid intruders and to protect the system, as the system is connected to the actuators automatically (SANG,

2006 apud GUBBI et al, 2013).

- **Addressing schemes**: For the Internet of Things to succeed, it is essential to have an exclusive competence to identify "things". The consequence is that billions of devices will be identified specifically and controlled remotely via the internet. To create a unique address for this function, some features are needed: (a) uniqueness; (b) reliability; (c) persistence and (d) scalability;
- **Data storage and analysis**: A huge amount of data is created as a result of this emerging technology. For this reason, data storage, ownership and validity are issues of concern. This data needs to be stored and used to monitor and act sensibly, which is why artificial intelligence algorithms need to be developed accordingly. Solutions based on cloud storage have become commonplace since 2012, and cloud-based analytics and visualization platforms are expected in the coming years;
- **Visualization**: for an IoT application, visualization is a fundamental process, because this is how the user interacts with the environment. With the progress of *touch* technology, the use of devices such as *smart tablets* and cell phones is natural and spontaneous. Visualization covers the identification of events and the visualization of unrefined data that is processed with information as the user needs it.

2.2 IOT SENSORS FOR DATA COLLECTION

One of the most important components for collecting data and presenting information that the IoT uses are sensors, which, depending on their application and purpose, can result in automated processes that can replace existing methods.

Hosain (2015) claims that, in addition to their location, devices that use the Internet of Things in M2M applications make use of sensors that collect information and help the device perform its functions. In addition, Hosain (2015) points out that sensors are circuits designed to meet integrated IoT/M2M applications because they have low-cost, miniaturized chips. The author Hosain (2015) highlights some of the most common sensors and their applications:

Table 2: Types of sensors and their applications [2]

Types of sensors and their IoT/M2M applications		
Types	**Description**	**Applications**
Acceleration sensors		

Accelerometers	They have the function of measuring acceleration (variation of the speed).	Measuring changes in speed and acceleration in cars; activating the *AirBag* in cars in the event of an accident.
Multiple axes of acceleration and sensitivity	Used to measure changes in vibration speed or in more than one dimension.	Two-axis sensors measure changes in two dimensions, and three-axis sensors are able to provide information in three dimensions.

[3] Source: Adapted from HOSAIN, 2015.

Types of sensors and their IoT/M2M applications		
Description	**Description**	**Applications**
Temperature sensors		
Silicon chip temperature sensor	The silicon chip sensor is used to measure the temperature range from -50 to 150°C and does not require strict calibration.	Used for measuring of temperature to help industrial process control.
Temperature sensor *Thermistor*	This type of sensor covers a measuring range of -100 to 450 °C, and is more accurate than the chip sensor. However, it is more complex to regulate in order to achieve the desired temperature accuracy.	

RTD (*Resistance Temperature Detectors*)	They are more fragile than silicon chip and thermistor sensors, and are more complex to use. However, their measurements are more accurate, but they are more expensive. They take measurements in the -250 to 900°C range.	
Temperature sensor *Thermocouple*	The most robust of the types presented. It measures a wider temperature range, from - 250 to 2000°C, allowing it to be applied to in processes chemicals and high-temperature furnaces in the semiconductor industry.	

Types of sensors and their IoT/M2M applications		
Applications		
Other types		
Light sensors	There is a wide variety of light sensors for a large number of applications. Infrared light sensors are another type of light sensor that is not visible to the human eye, however Perform their functions automatically when used as motion sensors.	Application of ambient light sensors (photocells inside the bulbs) that turn the bulbs on at dusk and off at dawn. Light detectors with sensitivity control adjusted by the owner.

Sensors MEMS (MicroElectroMechanical Systems)	They can transform physical movement into electrical signals, i.e. transform a movement into an electrical signal. into an electrical signal. It has low cost, excellent performance and high mechanical precision of the devices.	Check direction, pressure, movement; on smartphones, it works by automatically rotating the screen, display portrait landscape view by rotating the device. These chips are part of modern, high-quality smartphones and are also used to improve the user's or smartphone's location information.
Single switch sensors	Sensors that represent an open or closed state.	For example, door or window sensors in security systems. In homes where there are elderly people, the IoT and these sensors can detect when a compartment has been opened, such as medicine and food cabinets or the oven door.
Sensors Specialized	Sensors implanted in the body for a predetermined period of time. Research into these specialized sensors has been driven by MEMS sensors and semiconductors for the medical field.	Application in cardiac monitoring and vision correction; These sensors are being developed for implantation in pulmonary arteries to measure blood pressure and forward the information to a wireless device.

Chapter 4

4 CHAPTER - IOT APPLICATIONS

Given the positive view that IoT integrated with other technologies improves, assists and offers fruitful results in various areas, its applications cover an extensive range of segments and sub-areas, where the gains are reflected in improved services and user satisfaction.

As described by Vázquez (2013), objects connected to the internet generate a new business opportunity called *Servification*, based on the concept that the value of the physical object is passed on to the internet service, where these objects can undergo continuous improvements and adaptations in order to make them smarter and extend their useful life to the satisfaction of users.

In addition, Vázquez (2013) points out that the specific market for the Internet of Things has yet to be defined. However, its concepts can be used in various areas and businesses, such as logistics, connected furniture and appliances, monitoring systems for agriculture, smart clothing, toys, among others. This object + service approach could make people's lives easier, as forecasts suggest that 20,000 million or 50,000 million objects will be connected by the end of this decade.

Embedded systems, as reported by Kopetz (2011), can contribute to energy savings in various areas and to better use and efficiency of automotive fuels and energy used in household appliances. In this sense, the trend towards the installation of smart objects (IoT devices) is to help save energy and to monitor and reduce the energy supply according to consumption by installing smart meters.

Likewise, Rogers (2006) claims that the purpose of UbiComp technologies is to awaken reflection on the interaction between users and technologies, since there are benefits when this interaction is carried out to the right extent and can help to broaden human intelligence by improving the ability to learn, create, elaborate ideas, make decisions and solve problems.

With this in mind, the expansion of UbiComp technologies can be developed and customized for certain areas, such as agricultural production, environmental recovery or retail by companies, organizations or individuals.

According to Hosain (2015), the Internet of Things has a wide application in companies, which can be targeted at medical professionals as well as allowing companies to develop exclusive marketing campaigns for a single customer.

> One of the most basic uses of the IoT is to connect devices to the internet so that they can communicate their own status or their local environment. For example, an IoT/M2M device could be a temperature meter, a location sensor, a humidity measuring device, or an integrated circuit that checks vibration (HOSAIN, 2015, p. 8).

Kopetz (2011) highlights some areas in which the Internet of Things can be applied:

- **In the area of physical security and protection: IoT** devices can help detect counterfeit products or suspiciously reprocessed goods that are considered security risks in some sectors, such as airlines and the automotive industry;
- **The application of the Internet of Things in industry:** The Internet of Things applied in industry can help reduce costs in the diagnosis and maintenance of computerized industrial equipment, identify irregularities to prevent damage and

improve project safety;

- **In the medical field**: The Internet of Things can be widely applied in the medical field, such as controlling heart rate, blood pressure or administering the correct amount of medication via a smart implant, in the same way that smart *tags* can help patients with their medication. Another application would be to monitor the behavior of people with special needs and send messages if something unexpected happens.

In addition to these applications, Vázquez (2013) points out that the fundamentals of IoT can be applied to a city, where it would be possible to collect, store and analyze data to improve services provided by physical objects, such as urban furniture, pollution sensors, traffic signs, garbage collection or garden irrigation, to know the variations in temperature, wind, light, rain, among other activities. This stored data would represent knowledge about the city with the aim of analyzing the information to improve residents' quality of life.

In addition, Vázquez (2013) cites some examples of cities, including *Smart Santander (*Spain), *Smart City* Amsterdam (Netherlands) and *Songdo IBD* (South Korea), which are experimenting and deploying smart sensor networks to create conscious cities, known as smart cities.

According to Hosain (2015), some IoT devices are currently available for consumer use, such as:

- ***Nest* (owned by Google):** offers consumers a *Wi-Fi* connected thermostat that is self-configuring according to household habits, which can help through IoT and mechanisms that have made the whole house connected;
- ***Star* (*General Motors*):** One of the pioneers when it comes to connected vehicle services in the USA. Consumers of this product can count on navigation functions, on-board service and remote steering, among others;
- ***FitBit*:** It is part of a movement that began in 2000 with the principle of better personal understanding through data and technology. The purpose of this device is to monitor physical exercise via an internet connection, showing everything from step counts to heart rates,

The Internet of Things will influence various areas in which it can be used. According to Gubbi et al (2013), it is possible to classify applications into four domains, as described in Table 3:

Table 3: IoT application domains [3]

Domains	Description	Applied
Pe rsonal domestic	It uses WiFi networks, with high bandwidth and transfer rates, where individuals connected to this network can access information collected by the sensors (Haiyan et al, 2010; Nussbaum, 2006 apud Gubbi et al, 2013).	Sensors as an extension of the body to monitor elderly care (Haiyan et al, 2010; Nussbaum, 2006 apud Gubbi et al, 2013); Monitoring household appliances for energy saving and home management, among others (Darianian and Michael, 2008; Alkar and Buhur, 2005 apud Gubbi et al, 2013).
Enterprise	In this IoT domain, information is collected from internal networks of organizations emancipated by owners of these environments . Resulting in intelligent environments with unique configurations.	It can help with security , automation, temperature control and lighting, among others applications; Systems internal monitoring using sensors would replaced by wireless systems more versatile and specific adjustments.
Utilities	This application focuses on network information to improve consumer services rather than consumption.	Smart energy meter used by utility companies to manage costs in relation to profits.

Mobile	Technology available in mobile devices favors the implementation of the IoT as a new technology that will facilitate mobility and replace services provided by existing sensor systems. (Kumar et al, 2005 and Lin, Zito, Taylor, 2005 apud GUBBI et al, 2013)	Traffic, pollution and logistics management to make a more efficient transportation of goods (Kumar et al, 2005 and Lin, Zito, Taylor, 2005 apud GUBBI et al, 2013).

[3] Source: GARTNER, 2014 apud GARTNER, INC, 2015.

Gubbi et al (2013) point out that the arrival of the internet has been responsible for connecting people to an extent never seen before. The objects interconnected for the design of intelligent environments will be transformed according to the data relating to the connection of the devices.

(2014) the application domains of the interconnection of objects are shown in Figure 01 according to the scale of the impact of the data generated and users of this technological revolution.

Figure 1: Object linking application domains (GARTNER, INC. 2005).

In addition, Gubbi et al (2013) adds that in 2011, the number of interconnected devices exceeded the number of people on the planet, as currently 9 million devices are interconnected and by 2020 it is predicted that 24 billion devices will be interconnected.

The author adds that, according to the *GSMA* (*Global System for Mobile Communications Association*), which represents the interests of mobile operators around the world, this number of interconnected devices represents US$1.3 trillion in opportunities for mobile network operators in sectors such as health, automotive, utilities and electronics.

Table 4: Potential applications of the Internet of Things [4]

Groups	Area of Application		
Citizens	**Health**	**Emergency services, defense.**	**Crowd Monitoring**
	Screening; monitoring of patients; modeling disease spread and containment; temporal health status ; preventive information;	Personal remote monitoring (health, location); resource management and distribution, response planning	Crowd flow monitoring for emergency management ;use efficient public and retail spaces; workflow in commercial environments.
Transportation	**Traffic Management**	**Infrastructure monitoring**	
	Smart transportation through real-time traffic information and route optimization.	Sensors built into infrastructure to monitor structural fatigue and other maintenance ; Accident monitoring for incident management and emergency response coordination.	
Services	**Water**	**Building Management**	**Environment**
	Water quality , leakage, use, distribution, waste management.	Humidity control; energy consumption management.	Air and noise monitoring; watercourses; monitoring monitoring industry.

Potential applications of the Internet of Things identified by different discussion groups in the city of Melbourne are described in Table 4 above, where it is possible to identify the applications or use cases in an urban environment that benefit from WSN and the intelligent capacity of this resource (GUBBI et. al, 2013), out of context.

4.1 INTERNET OF THINGS AND APPS EDUCATIONAL

The new Information and Communication Technologies have led to the creation of a new concept called the internet of things, the main benefit of which is to develop

[4] Source: Adapted from GUBBI, 2013

intelligent multi-spatial, multi-functional and interconnected systems through networks and the internet. In this panorama, it is clear that, with the production of intelligent devices, the world wide web, which used to be, as the name suggests, made up only of computers, is now being integrated by intelligent things, machines and services.

These technological changes are rapidly affecting the way industrial processes are designed. It's no wonder that the Internet of Things is the idea behind the 4th industrial revolution.[a]

Experts also say that, in addition to industrial equipment and processes, professionals will also have to have expanded skills such as multidisciplinarity and solving complex problems that arise in the daily life of large cyber-physical systems (GORECKY, SCHMITT, et al., 2014).

One fact leads to another, because as mentioned above, the technologies that are emerging are based on the same concept as the IoT, and because they influenced factories by creating the 4th industrial revolution[a] , factories are now creating new demands for professional training because of this innovative organizational form of production processes. In addition, research focusing on "vocational education", which has always been widely studied, now gains a new set of possibilities through Remote Teaching/Learning Laboratories.

These environments are a step forward in complementing the teaching of physical sciences and technology in secondary and higher education courses, as they make it possible to interact remotely with real experiments, giving students flexibility in terms of both time and space. It is therefore a symbol of the "Internet of Things" in the field of education.

And because it plays a fundamental role in science and engineering education worldwide (HERADIO, TORRE, et al.), several applications have been developed, as in (MARCELINO et al., 2013), (MICHELS et al., 2016) and (ROSA et al., 2012). In addition to these, an application developed by the Federal University of Santa Catarina at the Araranguá Campus will be presented in the next subchapter, describing in detail an implementation of the Internet of Things in the development of an Educational Remote Laboratory.

4.2 DIDACTIC REMOTE PRESS

Scientific knowledge is actually an abstraction of reality. It is created by observing, studying, experimenting, analyzing, manipulating and transforming data.

Knowledge has various ways of being analyzed or expressed as symbols, equations, theories, principles, graphs, tables, images and so on. In other words, they are forms resulting from the interpretation of phenomena in practical life.

However, despite the fact that these representations of theory are already detailed in books and handouts, saving the learner a great deal of effort, what happens in practice is that simply reading is often not enough for the student to understand the knowledge. In fact, in the learning process, in order to build a mental model, it is essential to reflect on and understand the phenomenon from which the theory originated. For a lot of technical knowledge, students are not able to understand the theory well through explanations alone. It is important to visualize and interact with the phenomenon being studied in order to establish a link between theory and reality, thus establishing meaningful learning.

In this way, the role of remote experiments in learning is to help form mental models through observation and guided experimentation, in order to aid meaningful learning.

In technical electromechanical training and even in secondary school, one of the basic theories is *Hooke*'s Law in springs. This theory, like many others, can be demonstrated in practice through experimentation. However, there are often no viable resources or alternatives for demonstrating this phenomenon in practice.

One of the alternatives that has been implemented to aid educational processes and facilitate the learning of theory in practice are remote experiments. These devices are an example of how the idea of the Internet of Things can be present in all social and technological environments, making a relevant contribution. In this case, the aim of remote experiments is to provide experimentation, interactivity and study through personal devices with internet access, such as computers, smartphones and tablets. The Internet of Things is present, as it connects smart devices to teaching equipment.

As a way of spreading knowledge about the applications of the Internet of Things, the aim of this chapter is to demonstrate the development of the Didactic Press. This remote experiment was developed to help people understand *Hooke*'s Law by testing a helical spring and observing the relationship between the application of a force and the variation in its size. The main focus of this experiment is that it can only be accessed via the internet, using mobile devices or personal computers.

4.3 DEVELOPMENT OF THE DIDACTIC PRESS

The Didactic Press is an invention designed to be used to visualize the phenomenon of *Hooke*'s Law. To this end, this machine exerts a force on a helical spring in 4 positions and records the values of force (F) and variation in the size of the spring (Ax), providing data so that the student can observe and conclude the theory established by Robert *Hooke* in the elaboration of his law, which is summarized in equation (1).

Hooke's Law equation

$$F = k.\Delta x \qquad (1)$$

Where:

F = Force applied to the spring [N];

= change in spring size [m];

<= elastic constant of the spring [N/m]

The elastic constant (;-) is a property of each spring which reflects its resistance to elastic deformation. It is a characteristic that depends on the number of coils, the diameter of the coils, the diameter of the spring, the total size of the spring, the material and the heat treatment with which the spring was manufactured.

However, the characteristic that the student must be concerned with is only the phenomenon described by *Hooke*'s Law, which defines the existence of a certain linearity and proportionality of the variation in the size of the spring with the applied force (which must be within the yield strength of the metal). This proportionality is the key concept that is observed through the Didactic Press.

The student should conclude that, no matter what material the spring is made of, the important thing is to understand that the phenomenon of applying force to a spring occurs in all springs.

4.4 DETAILS OF THE PRESS CONSTRUCTION DIDACTICS

To understand the design of the "Didactic Press" remote experiment, Figure 2 provides an overview of the interconnections established between its main components. It can be seen that this composition is a set of systems that are geographically separated, but which together (via the Internet) form the overall structure of this experiment. This is understandable thanks to the idea of the Internet of Things, which turns objectives into ubiquitous systems.

Figure 2: Overview of the Didactic Press.

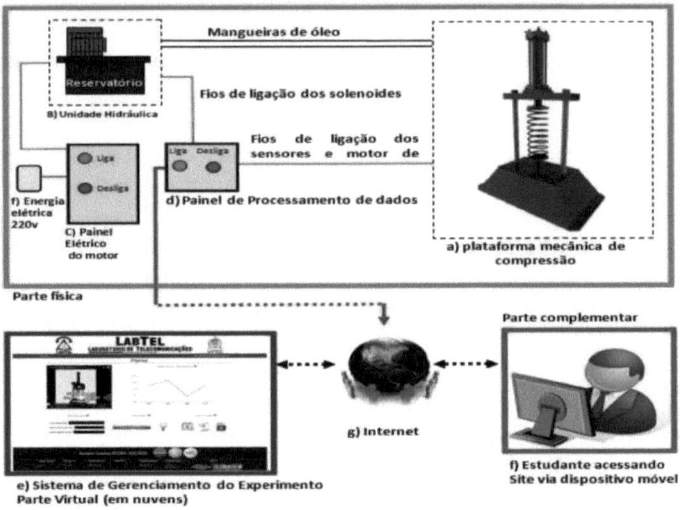

As shown in Figure 2, the Didactic Press experiment can be classified into 3 main parts: The physical part, the virtual part and the complementary part. The physical part is actually all the electromechanical components that make up the experiment itself. The so-called virtual part is the learning management site where the experiment's control panel is located. This site is hosted in the cloud on a third-party server. And the complementary part of the experiment is the internet access devices that access the site remotely. These are personal devices that have not been developed specifically for this experiment, nor have they been modified for this situation. They should only be directed to the specific address of the experiment.

Figure 3 shows an external view of the platform where the main components of the physical part are arranged. Figure 4 shows a schematic summary of the Didactic Press. The items are identified with letters and detailed below.

Figure 3: External view of the Didactic Press Platform.

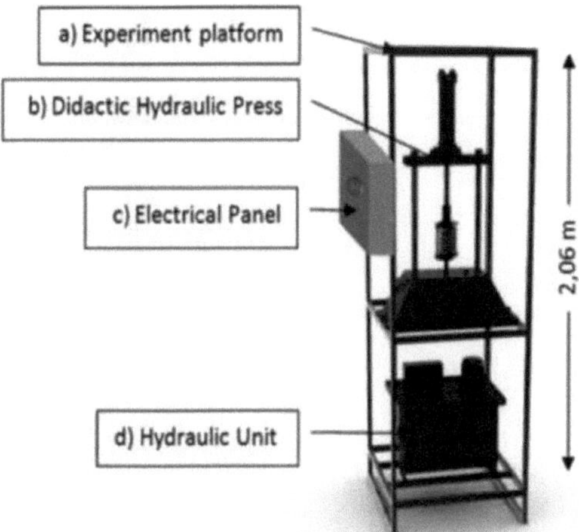

Figure 4: Main Components of the Physical Part.

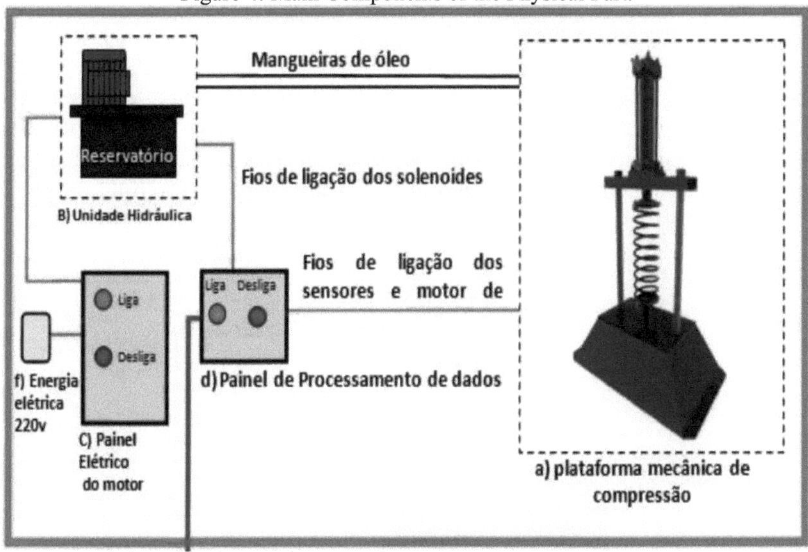

a) Mechanical compression platform

The mechanical compression platform is a key item in the remote experiment and is shown in Figure 5. The spring compression process takes place on it, as well as the collection of force data and the variation in the size of the spring. To capture the variation in spring size, a displacement sensor (c) was attached to the platform.

The force mechanism consists of a hydraulic cylinder system (a). The coil spring (f), common in cars. The other structures are made of steel, including the base.

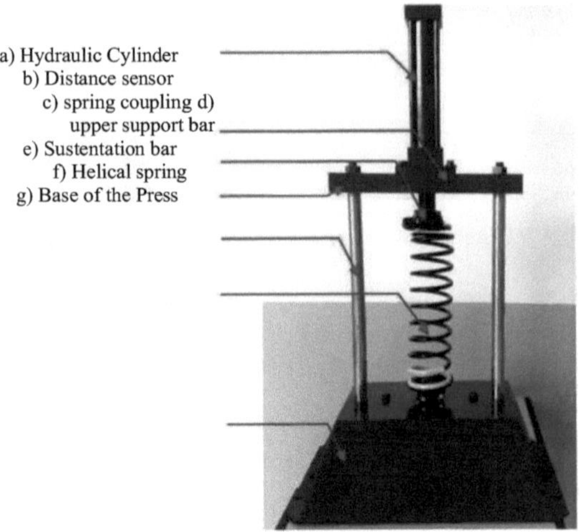

Figure 5: Detail of the mechanical compression platform.

The coil spring sits on a support, the shaft of which is attached to the inside of the platform base. This base is shown in detail in Figure 6, where the axis of the spring support is shown, which in turn sits on the load cell (shape I), whose function is to capture force data, which is exerted during the spring compression process.

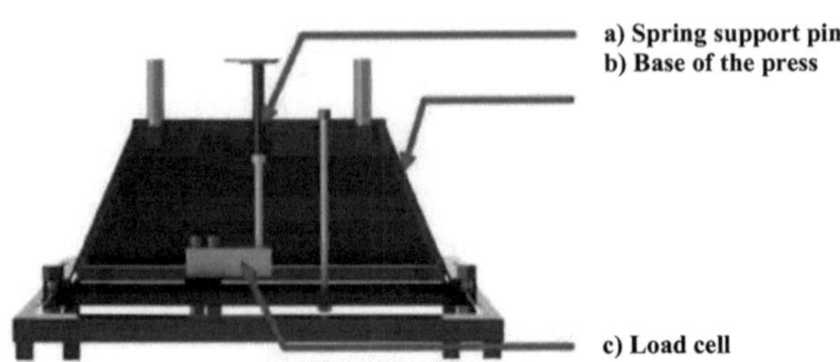

Figure 6: Base of the mechanical compression platform (sectional view).

b) Hydraulic Unit

Item B in Figure 4 represents the hydraulic unit illustrated in Figure 7, which is responsible for the power system that controls the hydraulic cylinder of the mechanical compression platform. Its function is to generate oil flow and control the direction of flow between the hoses, generating the forward and reverse movement of the hydraulic cylinder.

Figure 7: Hydraulic unit.

It consists of a hydraulic pump, an electric motor, a relief valve, a reservoir (80L) and a directional valve operated by a double solenoid. As a complementary component for controlling and driving the motor, there is an electrical panel, which is detailed in the next section.

The hydraulic circuit is shown in Figure 8.

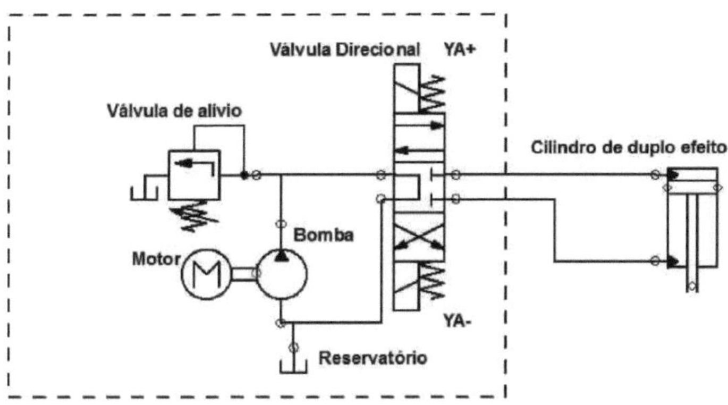

Figure 8: Hydraulic Circuit of the Hydraulic Unit.

c) Electrical Motor Control Panel

The motor drive electrical panel illustrated in Figure 9 is responsible for intermediating the control of the motor drive that touches the Hydraulic Unit's hydraulic pump.

Figure 9: Pump Motor Control Electrical Panel.

The main component is a frequency inverter, used to help adjust the motor's maximum speed and to facilitate external motor triggering by simple relays. In addition, this panel has safety devices such as thermal relays and circuit breakers that prevent damage from short circuits and overheating.

d) Data Processing Panel

The Data Processing panel, illustrated in Figure 10, is the main communication panel in the experiment. It is therefore a central panel, where all the components are connected to it.

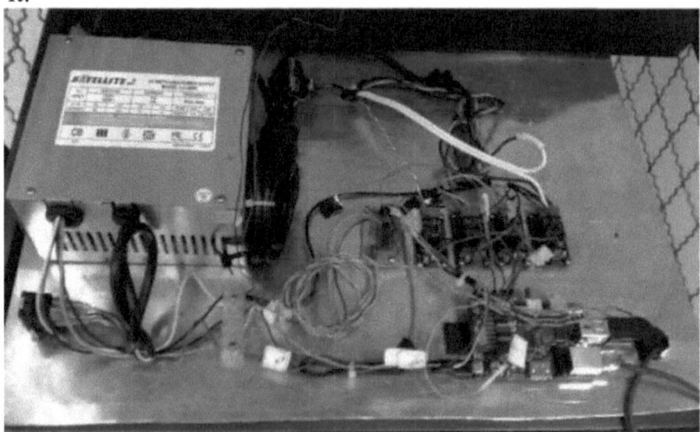

Figure 10: Data Processing Panel

The main component used in the panel is the *Raspberry Pi* illustrated in Figure 11. As well as exchanging data with the internet (server), its function is to control and monitor data. The *Raspberry Pi* is a minicomputer with *USB, RJ45, HDMI* ports and

GPIO (*General Purpose Input/Output*) pins for signal control. The *Raspberry Pi*'s main control and monitoring functions are detailed in Table 5.

Table 5: Raspberry Pi Control and Monitoring Items

Control Items	**Monitoring Items**
Hydraulic pump motor drive relays	Displacement sensor
Lamp drive relay	Force sensor (load cell)
Cylinder advance solenoid drive relay	Camera 1
Hydraulic cylinder return drive relay	Camera 2

This minicomputer, Figure 11, not only has a Linux operating system installed, but also programs to manage the devices and data, integrating the site with the experiment.

Figure 11 : *Raspberry Pi.*

As the *Raspberry Pi* does not have an analog data converter included, the *ABEletronics* ADC Pi *shield* shown in Figure 12 was installed to complete this fundamental function of collecting data from the analog force and displacement sensors.

Figure 12: Pi ADC signal converter.

The complete list of components in the data processing panel includes: 4 electromagnetic relays, a +12vdc +5vdc volt supply, a load cell signal amplifier, and the main component, the *Raspberry Pi*. The function of each component has been detailed in Table 6.

Table 6: Function of the Data Processing Panel Components

Item	**Function**
Raspberry Pi	Sensor data collection (force and displacement), internet communication, control of relays 1, 2, 3 and 4.
Source	Energizing the relays and the *Raspberry Pi*
Rele 1	Lamp activation
Rele 2	Inverter drive to control the hydraulic pump motor
Rele 3	Hydraulic cylinder advance solenoid drive (YA+)
Rele 4	Actuation of the hydraulic cylinder return solenoid (YA-).

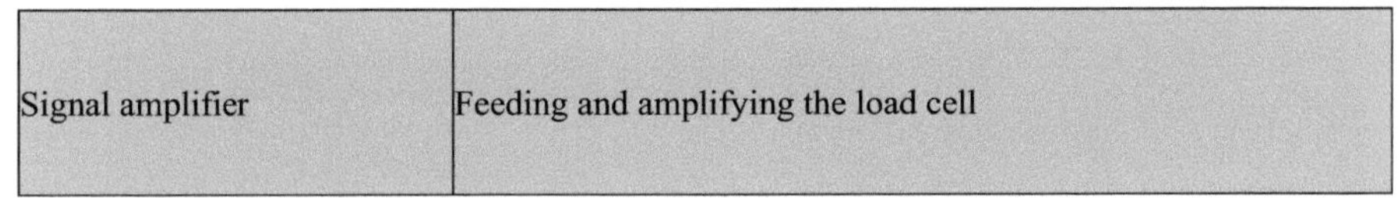

Signal amplifier	Feeding and amplifying the load cell

A detailed wiring diagram for each component can be found here. illustrated in Figure 13.

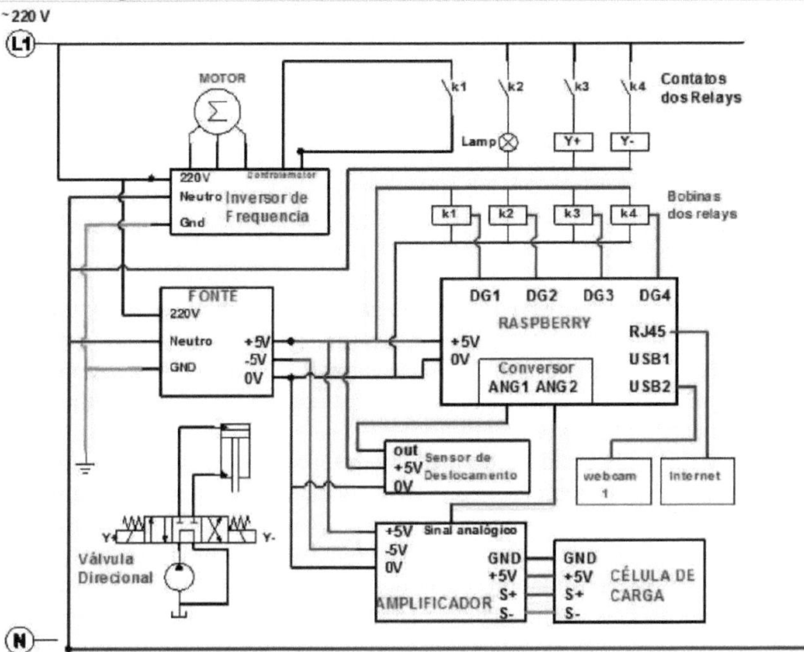

Figure 13: Data Processing Panel.

4.5 REMOTE EXPERIMENT WEBSITE

Considering the educational context of the project, it was decided that the press access control system would be based on a learning management system. With this in mind, we used the *Modular Object Oriented Dynamic Learning Environment* (*Moodle*), illustrated in Figure 14, because it is an environment that is widely used throughout the world and is open source, thus allowing it to be modified in order to achieve the layout and resources required for the remote press.

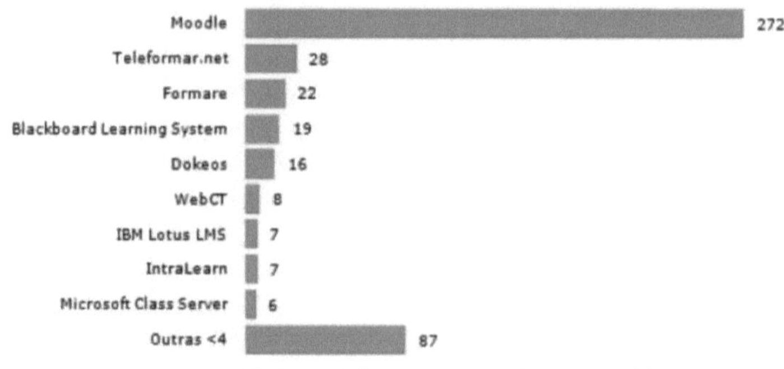

Figure 14: Comparison between e-Learning Platforms made by ED-ROM cited by (MARCELINO, 2010).

Moodle was created in 2011 by educator and computer scientist Martin Dougiamas, aimed at programmers and academics in education. The system allows a student or teacher to integrate with all the available resources in a simplified way, allowing them to connect even if there is differentiation by level within the software. Version 19 was used in this work.

Considering the tool already described, the remote lab's website was divided into three distinct parts: the remote experiment presentation page, the introductory environment for the experiment and the experiment page. This division was necessary in order to clearly separate the different environments within the laboratory website.

The remote press presentation page describes the didactic press, as well as presenting various aspects of the physical laboratory in which the press is positioned. This space is intended for presentations, i.e. it will not discuss details about the experiment and/or the experimentation process.

In the introductory environment, users of the control system have access to materials and exercises provided by a teacher in the field. In addition, users have access to various tools that allow user/teacher integration, exercises on the upcoming experiment and a tool for scheduling the experiment. This scheduling is necessary because resource control is required as there is only one physical press.

In the test environment, the input and output, control and video blocks are arranged on the page. The main aim of this page is to provide an environment where users can perform all the necessary controls, obtain output data and view the press in real time.

4.6 DEVELOPMENT OF THE PRESENTATION PAGE

Before developing the presentation page, the source code of one of the Moodle themes was modified in order to standardize the header and *footer of* all the pages. To do this, substantial changes were made to the *CSS (Cascading Style Sheets) of* the chosen theme. In addition, changes were made to *HTML (HyperText Markup Language)* and *PHP (Hypertext Preprocessor)*, which is object-oriented, in order to meet the standard defined by the authors.

After standardizing these spaces, the press presentation page was developed as shown in Figure 15. Different learning styles were taken into account. In addition to a textual approach, a visual approach was presented through the image of the experiment and a visual/audio approach through the *Voki* tool.

Figure 15: Press Presentation Page

Voki is a free service that allows you to create avatars, add audio, create a situation using a variety of backgrounds and then publish it. This tool becomes more of an attraction as it provides interaction between the environment and the user. Object-oriented *PHP, CSS and HTML* were again used to develop the *body of* this page. *Java Script was also* used.

4.7 DEVELOPMENT OF THE INTRODUCTORY ENVIRONMENT FOR THE EXPERIMENT

For the introductory environment of the experiment illustrated in Figure 16, all the modifications made to the website were made on the *Moodle* platform, i.e. there was no need to modify the source code. *Moodle*, being a learning management environment, already offers a large number of resources, including modules for providing teaching material, videos, exercises, among others.

This is a good thing, as anyone with minimal training in this learning tool will be able to manage the experiment material without having to rely on a developer.

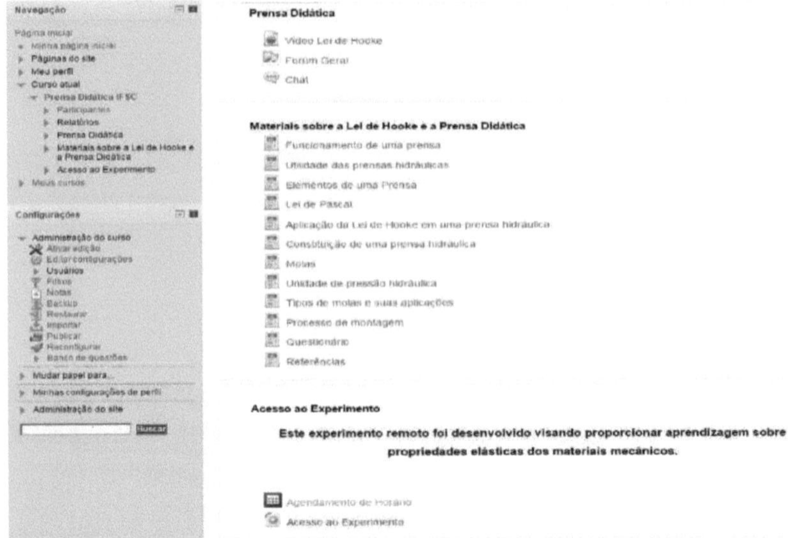

Figure 16: Introductory page to the experiment

In addition to all the modules already offered by *Moodle*, there is also the possibility of adding new features, such as the scheduling system, offered on the tool's support website. This has made it possible to enrich the page with different resources.

Considering the fact that the experiment under discussion was developed primarily for technical and high school students, the experiment page illustrated in Figure 17 was developed to create a simplified experiment process, where each of the five blocks within the body of the page clearly represents a step in the experiment process.

Figure 17: Experienced page

In the first block, at the top left of the page, you'll find a *streaming* video of the experiment. In the top right-hand position you'll find the *streaming* options, as well as a teaching video describing the process of running the press. In the bottom right position, the output data from the experiment is presented in the form of an interactive graph. At the bottom center, there are objects for setting the distances and starting the experiment. Finally, in the bottom left-hand corner, there is a simple calculator for calculating the spring constant.

To develop this page, in addition to using *PHP*, *CSS* and *HTML, we* used other resources such as *jquery*, *google charts*, *google fonts* and a *Facebook plugin*. These new resources helped to make the page more iterative.

4.8 EXPERIMENTING WITH THE DIDACTIC PRESS

As defined at the beginning of the chapter, the concept of the Didactic Press experiment is to perform a test on a coil spring in order to observe *Hooke*'s Law in practice. To carry out this experiment, we start by adjusting stop positions 3 and 4. As mentioned above, the data collection process only takes place at 4 stopping points (P1, P2, P3 and P4). The 1st and 4th points are pre-set at 1 cm and 22 cm respectively, leaving the student to adjust the distances of the 2nd and 3rd stopping points and then start the experiment. These processes are carried out on the website by items "f" and "h" in Figure 17. Table 7 lists the steps for carrying out the experiment.

Table 7: Experimental Procedures

Step	**Procedure**

1°	Open the Didactic Press control panel
2°	Release the start of the process (Start experiment button item "h" Figure 16)
3°	Set position 2 to between 1 and 12 cm. Note: as stop position 3 is still to be defined, it is necessary to leave a stop interval between points 3 and 4. You can enter these values via the numeric keypad or the scroll bar as shown in item "f" of Figure 16.
4°	Set the stopping point 3. according to step 3.
5°	Start experiment using the "start experiment" button. Note: This is the same button as "*set distance*".
6°	At the end of the experimentation process, the graph (as shown in Figure 17) will be displayed, based on the stop points set before the experiment.

Once the experiment has been completed, the student will be able to visualize the force data (F) and the variation in the size of the spring (Ax) at the points on the graph.

To view the data, hover the mouse pointer over the observed points. With this data, you can calculate the elastic constant (k) of the spring and visualize the linear relationship of *Hooke*'s law in practice.

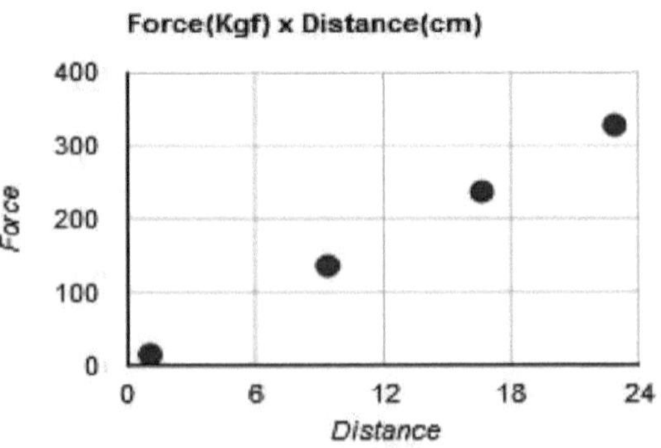

Figure 18: Pi ADC signal converter.

4.9 EXPERIMENT TEST RESULTS

To test the experiment, 42 trials were carried out with different stopping points, in order to check whether the data displayed by the graph would have the expected characteristics in accordance with *Hooke*'s law.

By overlaying the graphs generated, as shown in Figure 19, it is possible to conclude that the results of the experiments have the characteristics defined by *Hooke*'s law, as they show proportionality in the relationship "force x variation in spring size".

In addition, the illustration in Figure 18 also indicates that the results are repeatable. This makes the experiment suitable for didactic presentations.

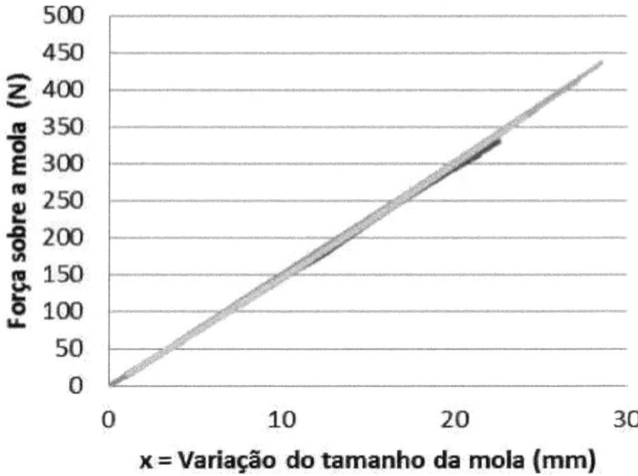

Figure 19: Overlay of the Results of the 42 Experiments.

Chapter 5

5 APPLICATION OF A REMOTE EXPERIMENT TO SUPPORT THE TEACHING OF HOOKE'S LAW TO TECHNICAL SCHOOL STUDENTS

The use of new information and communication technologies in education, especially IoT in conjunction with educational software, has provoked great interest in both classroom and distance learning. It is common knowledge that students of the technological generation find it difficult to attend traditional, monotonous classes. Thus, a remote laboratory can help in the learning of physical concepts, being an important resource in distance education courses that require practical classes, as well as face-to-face classes, making them more interactive and more dynamic. It can also help the learner independently of the classes and make it possible to carry out more complex and/or difficult-to-access experiments.

Today's students, young people who were born surrounded by technology, digital natives as we might call them, crave more innovative lessons that are less tiring than traditional ones. Especially when it comes to subjects like mathematics, physics and other sciences, subjects in which a lot of numbers and calculations are involved. It is therefore of the utmost importance to apply these concepts to the real world.

The research carried out used a remote laboratory that contains an experiment in the field of physics and is accessed through a learning management system.

The evaluation of this research was supported by the work of Marcelino (2010), which was based on the studies of Amigud, et al. (2002) and Nickerson, et al. (2005).

The evaluation based on a cognitive model sought to show that students have particular aspects and learn in different ways, which is why it is important to use the technologies used, as they are able to exploit all the individualities of the students, so that each one learns in the way that is easiest for them. The evaluation can be divided into four stages:

1ª Stage - The first assessment method consisted of applying the *VARK* questionnaire (Visual, Aural, Reader/Writer and Kinesthetic). This questionnaire was chosen because it is a concise instrument and has already been used in studies to assess which learning style the student prefers most. This questionnaire was created by Fleming & Mills (1992) and can be found online at http://www.vark-learn.com.

According to Fleming (1998), human beings have four learning channels:

I. **Visual:** These are people who learn more from information that is passed on visually, whether through pictures, graphs, films, chalk and blackboards, diagrams, slides.

II. **Auditory**: These are people who learn by listening. They learn more from the teacher's voice, spoken videos, tapes, sounds, discussions, seminars, debates, work presentations.

III. **Reading/Writing:** These people prefer written text in order to understand things better, summaries or schemes, arguments and discussions in written form, highlighting words.

IV. **Synaesthetic:** They learn by doing for themselves. They prefer real examples, practical work, technical visits, the use of metaphors in explanations, stories, handling objects involved in the subject.

This questionnaire seeks to show which of the student's characteristics are most evident. The use of this method is interesting both for students to find out which characteristics are most evident and thus exploit them to their advantage, and so that teachers can prepare their lessons by exploiting the various techniques and not using just one style that could harm a large number of students. Table 8 lists the different learning styles according to the *VARK* model and some teaching techniques that can be used.

Table 8: Relationship between Teaching Techniques and Learning Styles VARK [5]

Style Learning	**Teaching Techniques**
Visual	Lectures using blackboards, tape projections, Internet research, exercise solving, practical classes.
Hearing	Lectures, Seminars, Case Studies Developed in Groups, Lectures, Teaching in Small Groups, Debates.
Reading/Writing	Individual case studies, individual reading during and before class, development of summaries and essays.
Synesthetic	Seminars, exercise solving, practical classes, lectures by people in the field.

2ª Stage - The second stage consisted of conducting the lesson with the chosen remote experiment, which is accessed via the learning management system. First of all, the subject teacher gave a traditional lecture in the classroom to all the students. Then the class was divided into two groups, A and B. As the number of students in the class was odd (a total of seventeen), group A was left with nine (9) students and group B with eight (8) students.[13]

Only group A had access to the LABTEL6 portal and the remote experiment. Group B only had a traditional lesson from the teacher.

[5] Source: (MIRANDA, MIRANDA, MARIANO, 2014)

3ª Stage - The third stage consisted of applying exercises in the form of an assessment of the content explored for all the students, group A and B, to see what the impact was in terms of learning the content explored, whether there was any difference between the group that used the remote experiment and access to the material on the LABTEL portal and the group that only had the traditional lesson. In order to motivate the students to take part in this assessment, it was computed with a score determined by the teacher to be added to the students' marks.

Stage 4 - The last stage consisted of applying a questionnaire to assess student satisfaction with the technologies used. The aim was to identify the students' opinions on the use of these technologies, whether they approved of the use of the remote experiment, whether they found it easy to use, among other factors.

5.1 APPLICATION PROPOSAL

The proposal was applied using a remote experiment accessed via a learning management system (LMS) in a specific subject for technical courses.[14]

In order to find out whether there has been any real learning through the use of these new technologies applied in teaching, how satisfied the user is with the use of these technologies, among other questions, the application part of this work was based on a class of technical students from the Electromechanics course at the Federal Institute of Santa Catarina in Araranguá. This course was chosen because it has the subject Strength of Materials which studies the elastic properties of materials (*Hooke*'s law) and 17 (seventeen) students took part in this study.

5.2 LEARNING STYLES

Learning style is the method a person uses to acquire knowledge. It's not what a person learns, but how they behave while learning. Thus, knowing the learning style of students makes it easier to develop and use teaching methodologies and techniques that can be more motivating and meaningful in terms of learning effectiveness and thus have the possibility of generating much better results.

The first stage in applying the proposal was the *VARK* questionnaire, which surveyed the students' individual skills, which can be seen in Table 9.

Table 9: Questionnaire results [7]

	Visual	Hearing	Reading/Writing	Synesthetic
Number of Points	44	87	73	114

[7] Source: The Authors

The results of the questionnaire showed that most of the students assessed had

[14] LABTEL - Transdisciplinary Telecommunications Laboratory focused on research and development of data acquisition and monitoring solutions in the areas of New Information and Communication Technologies (NICT). It is located on the Mato Alto/Araranguá/Santa Catarina Campus at the Federal University of Santa Catarina/UFSC, situated in southern Brazil.

kinaesthetic characteristics, with a total score of 114 points, followed by auditory characteristics with 87, then reading/writing characteristics with 73 and the lowest of all the cognitive characteristics was visual with 44.

The proposed teaching environment, consisting of a traditional teacher's lesson and the use of a remote experiment together with materials from the LABTEL portal, aimed to reach all types of students, whether they were visual, auditory, readers/writers or synesthetic.

In the teacher's traditional class, the students' visual characteristics were explored, using the blackboard and slides. The auditory characteristic was also explored in the teacher's class, as voice was used to explain the content and through the videos on the subject made available on the LABTEL portal.

The Reading/Writing and Synesthetic characteristics were more evident in the remote experiment. All the theory taught in the teacher's traditional class was available to the students on the LABTEL portal, so the students with the Reading/Writing characteristic could access the portal and read all the content through the texts and carry out calculations with the data obtained from the experiment.

The remote experiment was designed to satisfy synesthetes. By accessing and carrying out the remote experiment, the student can prove the theory presented in practice. Table 10 shows in simplified form the learning styles used in the proposal.

Table 10: Learning Styles Used [8]

Learning Style	**Teaching Techniques Used**
Visual	Lectures using blackboards, tape projections, Internet research, exercise solving, practical classes.
Hearing	Lectures, Seminars, Case Studies Developed in Groups, Lectures, Teaching in Small Groups, Debates.
Reading/Writing	Individual case studies, individual reading during and before class, development of summaries and essays.

Synesthetic	Seminars, exercise solving, practical classes, lectures by people in the field.

[7] Source: The Authors

5.3 APPLICATION OF THE PROPOSED MODEL

The second stage of the application was the implementation of the proposed teaching model, which aimed to exploit all the skills. Firstly, the subject teacher gave his traditional lesson on the elastic properties of materials and *Hooke*'s law to the whole class. Then the class was divided into two groups, A and B, with group A having 9 (nine) students and group B 8 (eight) students. Group A had access to the IFSC Didactic Press course on the LABTEL portal. The students had time to use all the resources and materials provided and were able to carry out the remote experiment and make the calculations relevant to the experiment. According to the report, "*all the students were very excited about the lesson using the experiment and they all paid close attention and were very involved in the moment. Before this class, the teacher of the subject had already commented that he thought the students would like it, as they were complaining about the extremely theoretical classes full of calculations*"

In the third stage, the aim was to assess learning through the use of this resource. To this end, all the students, groups A and B, took an assessment in the form of exercises and questions about the content covered. The motivation for answering this assessment came in the form of an extra mark that would be scored by the subject teacher and added to the participants' marks. The students completed the assessment individually and had a set amount of time to complete the exercises. The purpose of applying the exercises to all the students was to assess whether there was any difference between the students in group A and group B.

The last stage consisted of administering a questionnaire to the students in group A about their satisfaction with the resources used. The questionnaire was applied on a handwritten sheet and only the students in group A answered it, as they were the ones who used the remote experiment and other resources and this questionnaire aimed to find out their opinion of the technologies used.

5.4 RESULTS

Information and Communication Technologies are increasingly being used and disseminated around the world. Education is increasingly using technological devices to create learning situations that are in keeping with today's world and the demands of the knowledge society. However, the great force of transformation will be the students themselves, who are becoming increasingly familiar with the new technologies and as soon as one teacher begins to adopt new techniques, the others will be held to account.

In the research carried out, the proposal applied was evaluated in 2 (two) items. The first item was an evaluation containing 10 (ten) exercises prepared by the teacher of the subject, which were based on the elastic properties of materials (*Hooke*'s Law).

Remember that the students were divided into two groups, A and B, so in this first assessment item, all the students in both groups did the exercises. The result of the first assessment item, which can be seen in Table 11, shows the individual scores of each student, the average of group A and group B and the standard deviations. In conclusion, group A had a better result compared to group B, obtaining an average of 8.15 against 5.56. Group A was the group in which the students had access to the LABTEL portal and the remote experiment, so it showed an advantage over the group that didn't use the same resources.

Table 11: Exercise Evaluation Results [9]

	Group A	**Notes**	**Group B**	**Notes**
	Student A	8	Student J	8
	Student B	6,8	Student K	6,5
	Student C	6,5	Student L	7,2
	Student D	7	Student M	4,5
	Student E	8,5	Student N	2
	Student F	8,5	Student O	5,8
	Student G	8,5	Student P	4,5
	Student H	9,5	Student Q	6
	Student I	10		
Average		8,15		5,56
Standard Deviation		1,13		1,75

[9] Source: The Authors

The second evaluation item was a questionnaire on student satisfaction with the tools used. The application of this questionnaire and the questions asked in it were based on the work of Marcelino (2010), who according to the author "sought to discover the satisfaction of using the new tools, the ease of use, the problems with the technology. We tried to extract the student's feelings about the new proposal" (MARCELINO, 2010, p.100). Table 12 shows the general aspects evaluated, showing the average and standard deviation for each item.

Table 12: Results for the General Aspects of the Proposal [10]

General aspects	Average	Standard Deviation
General satisfaction	8,89	0,99
Clarity of Instructions	9,45	0,68
Learning support	9,11	0,74

[7] Source: The Authors

As you can see, the item with the highest score was the clarity of the instructions. As the proposal was something innovative for the students, we tried to make it very clear how the system worked, pointing out that the LABTEL portal uses the *Moodle* tool to manage its site, which allows the resources to be intuitive and clear. The experiment also included several written and video explanations. The overall satisfaction score was good, showing the students' acceptance. The item evaluating the process of aiding teaching and learning scored highly, corroborating the first evaluation item (in which group A had a good average), which shows that the students really did have an aid to learning with the use of the tools.

Table 12: Results of the Remote Experiment [11]

Remote Experiment	Average	Standard Deviation
User satisfaction	9	0,94
Ease of use	9,67	0,67

Objective and obvious system	9,12	1,05
Help with Practical Subjects	9,20	0,78

[7] Source: The Authors

Table 13 shows the results of the questions to evaluate the use of the remote experiment itself. We can see that all the items scored highly. The items ease of use and objective and obvious system show how intuitive and easy to use the remote experiment system is, making the students less afraid and more excited about using the tool. The item help in practical subjects represents very well the importance given by the students to the use of remote experiments in teaching.

Table 14: Results of using the LABTEL Portal [12]

LABTEL Portal (*Moodle*)	Average	Standard Deviation
Satisfaction with using the *Moodle* system	9,33	0,82
Use as a complement to classroom lessons	9,55	0,50

[7] Source: The Authors

Table 14 refers to the LABTEL portal, which uses *Moodle* software to manage resources. As the IFSC does not use any Learning Management System or Virtual Learning Environment to manage and make available the content and materials worked on in the classroom, in the case of this subject, the students only have notes taken in the classroom or materials that the teacher sends via each student's personal e-mail. We therefore sought to identify how satisfied students were with using the LABTEL system, which provided materials and other learning support resources. The score of 9.33 for this item made it clear that the students approved of using a system that brings everything together in one place, making the student's life much more organized. The second item identified their approval of the use of a system such as the one used to complement face-to-face classes, given that they don't have one. The last question in the satisfaction questionnaire was an open-ended question, in which students could write comments, criticisms and suggestions in general. Only three students gave their opinion, but all the comments were positive. Table 14 shows what the students wrote.

Table 15: Table of comments made by students [15]

Very good and very interesting work, a good new way of transforming dull lessons into simple ones that make people more interested in learning.
Congratulations on the initiative and the practical experiment.
Classes like this help to break up the monotony of theoretical classes like this one.

Table 16 is a correlational table involving the assessment methodologies used. It involves the students, the marks for the assessment exercise, the questionnaire assessing student satisfaction with the technologies that were applied and the *VARK* questionnaire.

Table 16: Correlation Table between Evaluation Methodologies [14]

				Student Satisfaction Evaluation Questionnaire						***VARK***			
			Evaluation Exercise	Overall satisfaction	Help with learning	Remote Experiment	Instructions	Ease of use	Objectivity and obviousness	Visual	Hearing	Reading/Writing	Synesthetic
GROUP A	1	Student A	8,0	10	8	10	10	10	10	3	4	4	5
	2	Student B	6,8	9	10	9	9	10	9	5	6	5	7
	3	Student C	6,5	10	10	8	10	10	10	2	4	7	5

[15] Source: The authors.

	4	Student D	7,0	9	9	10	10	10	10	3	7	7	6
	5	Student E	8,5	8	9	8	9	10	9	3	2	3	8
	6	Student F	8,5	10	10	10	10	10	10	5	7	3	10
	7	Student G	8,5	7	9	8	8	8	7	2	8	4	2
	8	Student H	9,0	9	9	10	10	10	10	1	5	2	10
	9	Student I	10	8	8	8	9	9	8	0	5	3	8
GROUP B	10	Student J	8,0	-	-	-	-	-	-	1	4	4	7
	11	Student K	6,5	-	-	-	-	-	-	6	10	4	8
	12	Student L	7,2	-	-	-	-	-	-	3	6	2	5
	13	Student M	4,5	-	-	-	-	-	-	2	4	4	7
	14	Student N	2,0	-	-	-	-	-	-	4	2	5	7
	15	Student O	5,8	-	-	-	-	-	-	1	7	6	9

	16	Student P	4,5	-	-	-	-	-	-	2	5	6	3
	17	Student Q	6,0	-	-	-	-	-	-	1	1	4	7
	Average		6,9	8,9	9,1	9,0	9,5	9,7	9,1	2,6	5,1	4,3	6,7
	Deviation		1,9	1	0,7	0,9	0,7	0,7	1,1	1,6	2,3	1,5	2,1

[7] Source: The Authors

At the end of the table are the final averages and their respective standard deviations. Several observations can be made from this table. Starting with the marks for the assessment exercise, it can be seen that the four lowest marks (4.5, 2.0, 5.8 and 4.5) came from group B, whose students did not have access to the LABTEL portal and the remote experiment. Another observation to be made is that the students who obtained low scores (Student M - 4.5; Student N - 2.0 and Student O - 4.5) in the *VARK* result showed greater synesthetic characteristics. It should also be noted that the five (5) highest scores (8.5; 8.5; 8.5; 9.0; 10.0) came from group A, which had access to the LABTEL portal and carried out the remote experiment. Four of these students (Student E - 8.5; Student F -8.5; Student H - 9.0 and Student I -10.0) showed synaesthetic characteristics. These results lead us to conclude that the inclusion of new information and communication technologies, such as remote experiments, is an alternative that can bring many benefits to teaching and learning.

The *VARK* questionnaire showed that in 65% of all students, the characteristic that stands out the most is kinaesthetic, the rest are multimodal, alternating between values for auditory and reading/writing characteristics. No student was visual.

Regarding the questionnaire evaluating student satisfaction with the technologies that were applied, all the scores were high, showing that in general they enjoyed the experience, i.e. they approved the use of remote experiments and *SGA S* as an aid to learning.

The IoT makes it possible to increase efficiency by integrating devices and utensils connected to the Internet, helping students and teachers in the teaching and learning process.

It shows that the IoT has the potential in education to provide remote access to educational tools in order to allow students to clarify doubts while studying in remote locations, as well as to share information with their teachers and classmates on specific platforms.

Using the potential of the IoT in education also involves developing models that can provide students with customized learning experiences for each type of need.

The role of IoT is to provide a teaching and learning environment that improves control and responsiveness in a more collaborative education. It is said that as well as changing people's daily lives, the IoT, with its constant and rapid evolution, will change the way students learn, how they interact and collaborate with each other and how teachers teach. These transformations give us a glimpse of what the future of education will be like and how schools, universities and educational institutions need to be prepared to adapt and withstand it all.

The conclusion is that the IoT will be largely responsible for various teaching and learning solutions that will be applied in different areas of knowledge, thus bringing a better quality of life to all people.

BIBLIOGRAPHICAL REFERENCES

[1] FLEMING, Neil; MILLS, Collen. **Not Another Inventory, Rather a Catalyst for Reflection. The Professional and Organizational Development Network in Higher Education**. Centerbury, v. 11, p.137-155, January 1992. Available in: <http://digitalcommons.unl.edu/cgi/viewcontent.cgi?article=1245&context=pod improveacad>. Accessed on: Aug. 11, 2016.

[2] GARTNER, Inc. **Gartner's 2014 Hype Cycle for Emerging Technologies Maps the Journey to Digital Business**. 2014. Available at: <http://www.gartner.com/newsroom/id/2819918>. Accessed on: May 23, 2016.

[3] Gartner's 2015. **Hype Cycle for Emerging Technologies Maps the Journeyto DigitalBusiness .** 2015.Available in: <http://www.gartner.com/newsroom/id/3114217>. Accessed on: 07 Jun. 2016.

[4] Gartner Says. 4**.9 Billion Connected "Things" Will Be in Use in 2015: In 2020, 25 Billion Connected "Things" Will Be in Use**. 2014. Available at: <http://www.gartner.com/newsroom/id/2905717>. Accessed on: 07 Jun. 2016.

[5] GORECKY, D. et al. **Human-machine-interaction in the industry 4.0 era. Industrial Informatics (INDIN)**. [S.l.]: 12th IEEE International Conference on. 2014. p. 289-294.

[6] GUBBI, Jayavardhana et al. **Internet of Things (IoT): A vision, architectural elements, and future directions.** Future Generation Computer Systems, [s.l.], v. 29, n. 7, p.1645-1660, sep. 2013. Elsevier BV. http://dx.doi.org/10.1016/j.future.2013.01.010.

[7] HERADIO, Ruben et al. **Virtual and remote labs in education: A bibliometric analysis**. Computers & Education, [s.l.], v. 98, p.14-38, July 2016. Elsevier BV. http://dx.doi.org/10.1016/j.compedu.2016.03.010.

[8] HOSAIN, Syed Zaeem. **The Definitive Guide: The Internet of Things for Business.** Santa Clara, Ca: Aeris Communications, Inc., 2015. 131 p. Available at: <http://info.aeris.com/iotguide>. Accessed on: May 28, 2016.

[9] KOPETZ, Hermann. **Real-Time Systems Series: Design Principles for Distributed Embedded Applications**. 2. ed. Austria: Springer Science & Business Media, 2011. P.307-322.

[10] MARCELINO, Roderval. **Virtual Learning Environment Integrated with 3D Virtual World and Remote Experiment Applied to the Strength of Materials Theme.** 2010. 128 f. Thesis (Doctorate in Engineering) - Postgraduate Program in Mining, Metallurgical and Materials Engineering - Federal University of Rio Grande do Sul, Porto Alegre, 2010.

[11] MARCELINO, Roderval et al. **Remote Educational Experiment Applied To Electrical Engineering.** International Journal of Online Engineering (ijoe), [s.l.], v. 9, n. 1, p.22-25, February 11, 2013. International Association of Online Engineering (IAOE). http://dx.doi.org/10.3991/ijoe.v9i1.2298.

[12] MICHELS, L. Boeira et al. **Remote Compression Test Machine for Experimental Teaching of Mechanical Forming**. International Journal of Online Engineering (ijoe), [s.l.], v. 12, n. 04, p.20-22, 28 abr. 2016. International Association of Online Engineering (IAOE). http://dx.doi.org/10.3991/ijoe.v12i04.5085.

[13] ROGERS, Yvonne. **Moving on from Weiser's Vision of Calm Computing: Engaging UbiComp Experiences.** Lecture Notes in Computer Science, [s.l.], p.404-421, 2006. Springer Science + Business Media. http://dx.doi.org/10.1007/11853565_24.

[14] ROSA, M. et al. **Remote monitoring system to measure the temperature of small wind power using Power Line Communication - PLC technology**. 2012 9th International Conference on Remote Engineering and Virtual Instrumentation (rev), [s.l.], p.1-5, jul. 2012. Institute of Electrical & Electronics Engineers (IEEE). http://dx.doi.org/10.1109/rev.2012.6293147.

[15] VÁZQUEZ, Ignacio Juan: **19 fundamental essays on how internet is changing our lives**. Portugal: BBVAOpenMind, 2013. 471 p. Available at:<https://www.bbvaopenmind.com/wp-content/uploads/2014/04/BBVA-OpenMind-libro-Cambio-19-ensayos-fundamentales-sobre-cómo-internet-está-cambiando-nuestras-vidas-Tecnología- Interent-Innovación.pdf>. Accessed on: May 22, 2016.

[16] GRUBER, Vilson. et al. **Compression Testing Machine in the Age of the 4ª Industrial Revolution.** Ferramental (Curitiba), v. September, p. 17-23, 2016.

[17] GRUBER, Vilson. et al. **The Internet of Things applied to the concept of energy efficiency: a quantitative-qualitative analysis of the state-of-the-art literature**. AtoZ: new practices in information and knowledge , v. 5, p. 80-90, 2017.

Printed by Books on Demand GmbH, Norderstedt / Germany